Get started in electronics: and basic concepts

Arianne Luna

Essential concepts

This book will be a first step for those who wish to begin to transit the learning and practice of electronics.

We can define electronics as a branch of physics and engineering that analyzes and employs systems based on the conduction and control of electricity for specific purposes, all through the creation of an electrical circuit. But what do we mean when we talk about voltage, resistances or current? Before we focus on these concepts, we must start with the primary element: electricity.

What is electricity?

Electricity is everywhere. Many times it originates in natural phenomena that are difficult to perceive, because the process is generated at the atomic level, small particles that we cannot capture with the naked eye. We could define it as a movement or flow of electrons through a material that conducts them.

Today we need electricity to run most of our technology. The great advantage of electricity is that we can convert it into other forms of energy: heat, light, movement...

To describe electricity we must start from the atom. The atom, as we know, has negatively charged particles called Electrons and positively charged particles called Protons. The electrons orbit around the nucleus of the atom.

We can observe how the atom has several "layers", ranging from the closest to the nucleus to the farthest, where electrons circulate. Now, this is where we enter into the concept of the elements called "metals" and "non-metals". You probably remember them when you studied the periodic table, don't you? The particularity of the elements called Metals is that they can exchange electrons between their atoms. In other words, the electrons that circulate through the outer layers can pass from one atom to another. Electrons that can circulate between atoms are called "free electrons". The orbit we are interested in is the last, or most external. The electrons present there are called "valence electrons".

The conductive metals are those that facilitate the current circulation. The best metals and the most used for conducting current are copper, silver, gold and aluminum, as they all have less than four valence electrons. Elements with a valence electron are usually the best conductors (copper, for example).

When many electrons move through the atoms of a material, they create an electrical flow.

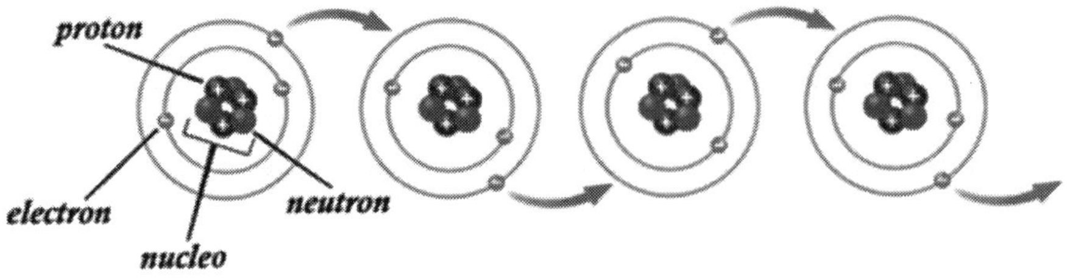

Now let's add the concept of Polarity. This is based on the fact that negatively charged elements are attracted by positively charged (opposite) elements. On the other hand, elements with equal charge (positive with positive, for example) are repelled.

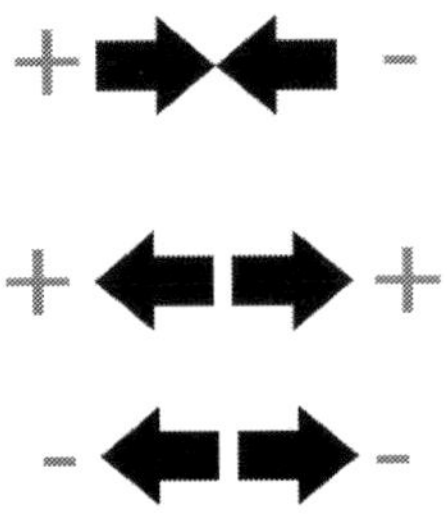

Current

Let's take the example of a cable, the inside of which is made of copper:

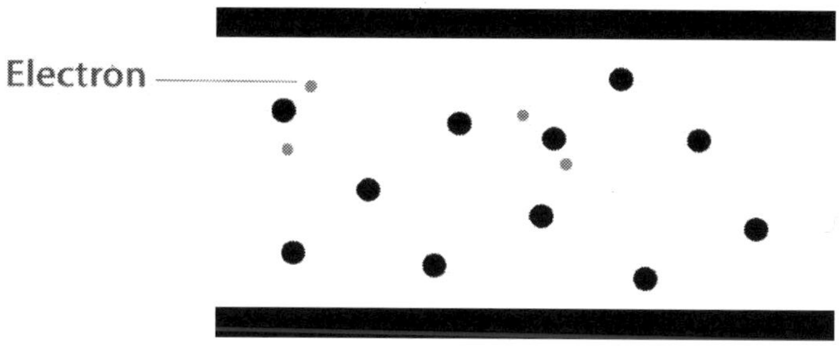

Here we see that the electrons are at rest. Since there is no force to move them strong enough to generate an electric current. Knowing that the electrons, with their negative charge, will move to any place that contains a positive charge, and in turn, will move away from an area of negative charge, we can make the current flow, how? Place a positive charge on one end of the cable and a negative charge on the other end.

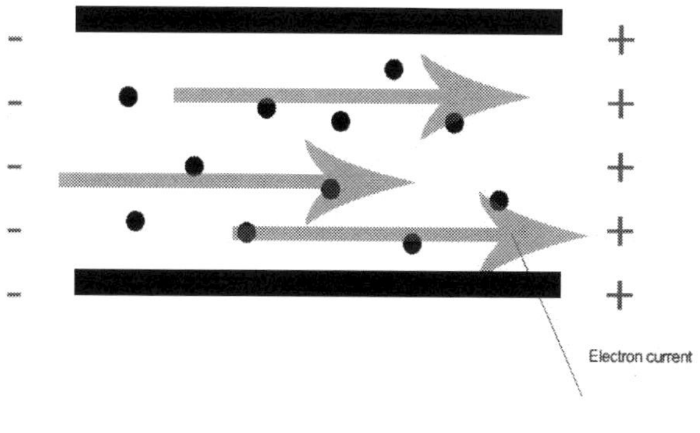

In this way the electrons pass from one atom to another and all travel in one direction: towards positive charge. This flow of electrons (amount of them circulating in a given medium), has as its name Current or Intensity (represented by the letter I), and is measured in Amperes (A).

How can we increase the electron current? So very simple, increasing the positive and negative charge, which we had placed on both ends of the cable. In this way, electrons that had not been mobilized before cannot now resist the attraction of positive charges, thus increasing the current.

Voltage

The force produced by negative and positive charges, and therefore the force or tension with which electrons move, is called Voltage, and is measured in Volts (V). The more voltage there is, the more current there is, because the more electrons move.

For electricity to be generated, a voltage source is needed. That is represented in the figure above by both sides (negative and positive charges) that are responsible for stimulating electrons. Voltage sources can be batteries or generators and are represented by the following symbol:

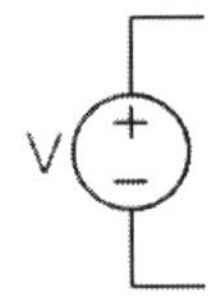

Resistance

Resistance can be defined as the opposition of a material to the passage of the current. The greater the resistance, the more it will cost for the current to pass through a medium. This is directly related to conductivity. A conductive material such as copper offers very little resistance, and a material such as plastic (insulation), has a high resistance to the passage of current. Conductive materials that offer little resistance have more free electrons, while insulators have little or no free electrons. Resistance is usually represented in diagrams with the following symbol:

Resistors in circuits are usually used to regulate current or voltage, according to the needs of each element of the circuit. The excess energy (the energy that does not pass) is released in the form of heat. Hence many electric stoves and toasters are basically resistors, as the current passing through certain filaments meets a resistant medium and, when electrons struggle to pass through them, they release energy in the form of heat. Resistance is represented by the letter R and is measured in Ohms (Ω).

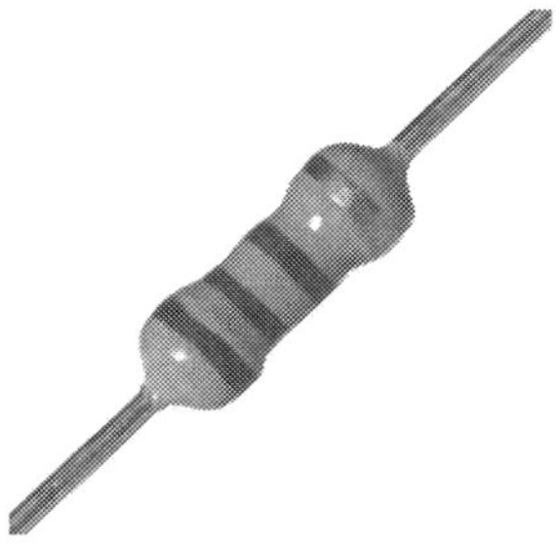

Figure 1: Typical resistance used in circuits

Power

Power is the relationship between energy and time. In simple words, it is the amount of current consumed by an element in a given period of time. For example, we have a car whose engine is running but in a resting state. On the other hand we have a car accelerating at full speed. The engine of the second will consume more energy per second than the first, so it will have more power. Power is measured in Watts (W). To calculate the power of any element is used the formula: voltage (volts) per intensity / current (amps).

With the Watts triangle we can clear the value of any of these three elements: current, power and voltage:

Voltage = Power / Current
Intensity = Power / Voltage
Power = Voltage x Current

You're wondering how we could calculate the Resistance. The Law of Ohm is used for this.

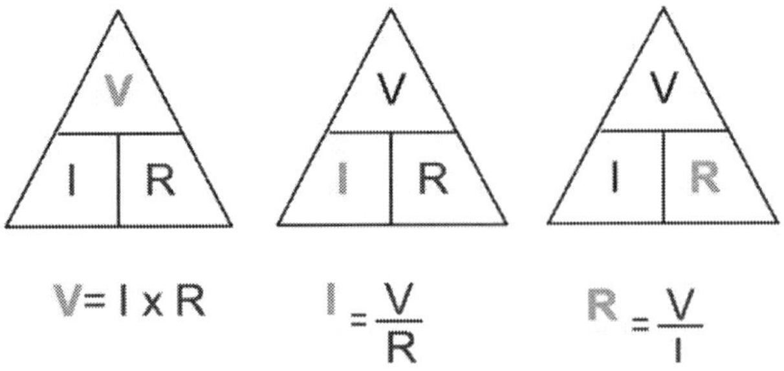

By holding two of the three data we will be able to clear a third.

Practical example

To better understand all concepts using an analogy, we can say the following: Imagine you have a faucet or water faucet. You connect a hose to this faucet. You notice that the water comes out with quite a lot of force (pressure). The force with which the water comes out of the tap would be equivalent to the Voltage of a circuit. The speed that the water uses to circulate through the hose would be the Current, we could also compare it with the amount of water that circulates through the hose in a certain period of time.

Now, how could we represent resistance here? Very easy. You realize that the hose has too strong and thick a jet, and you want to get a thinner and weaker jet of water. The hose has too large a diameter, so at the end of the hose you add a joint to join another smaller diameter hose. Now the amount of water coming out of the end decreases.

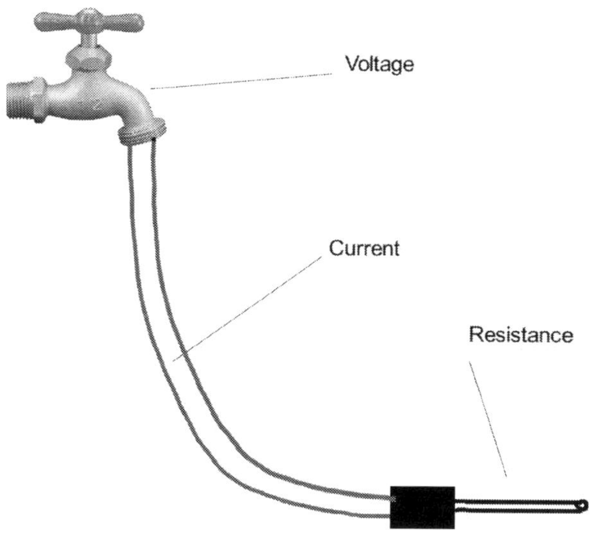

Alternating current and direct current

In electronics we can find two types of currents. On the one hand we find alternating current, which is present in external electrical networks, as in the outlets that we find in any house. Direct current is the kind of current found within internal circuits, for example within devices such as computers or mobile phones.

Alternating current is that whose electron flow varies constantly, in the form of regular cycles. That is to say, the current can adopt negative and positive parameters cyclically, it can flow in several directions. The advantage of alternating current is that it does not have large losses due to distance, unlike continuous current, which is prone to lose intensity.

On the other hand, **direct current** is stable, has no cycles or variations in the flow of electrons, and flows only in one direction, produces less noise and interference and is much more malleable and modifiable.

Fiaure 2: Direct current

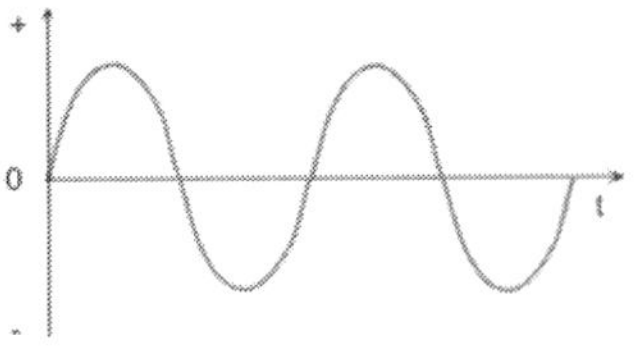

Figure 3: Alternating current

Electrical Circuits

An electrical circuit could be defined as a set of elements connected to each other, through which a current used for certain purposes or results circulates. In a circuit we generally use energy to convert it into other forms of energy, or convert certain types of energy signals into other types of signals.

A primary element of the circuit is the **Voltage Source**, which is responsible for supplying current through the entire circuit.

A second essential element is the **Conductor**. It is responsible for conducting and carrying the current through the path we want to trace in the circuit and all its components.

A third essential element is a **Power Receiver**: it can be a lamp, a motor, a loudspeaker... any element that makes use of current to achieve a specific purpose.

Types of circuits

There are several types of circuits: simple, series, parallel and mixed. Next we'll see an image of a simple circuit, consisting of only one receiver:

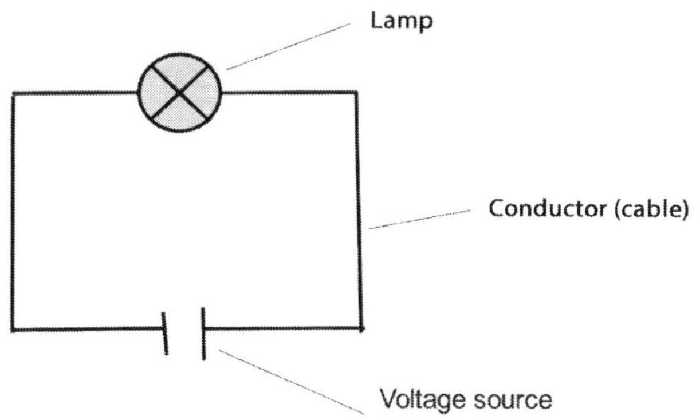

Series circuit

This is a circuit where we connect several receivers in series, one after the other:

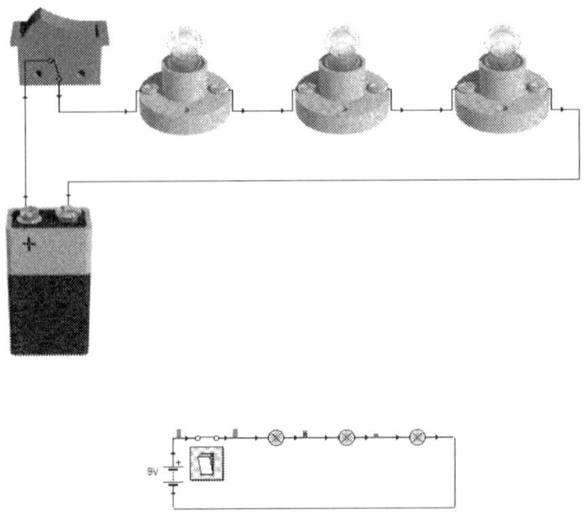

Figure 4: Series circuit. Made by www.edu.xunta.es CC-BY-SA License.

Parallel circuit

It is a question of making a connection whose receiving elements are independent and are not one after the other, but are fed by a parallel current.

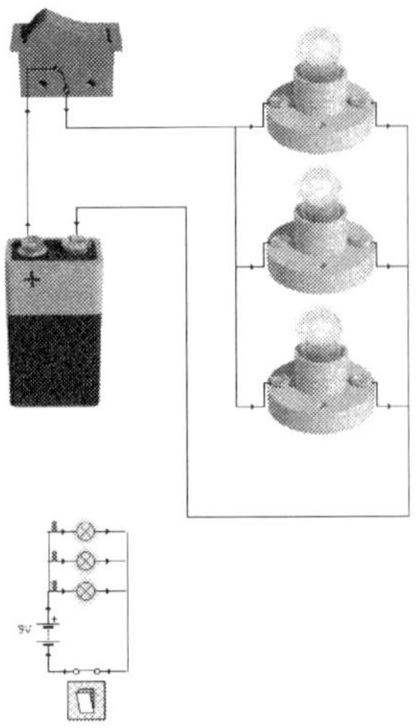

Figure 5: In parallel. Made by www.edu.xunta.es. CC-BY-SA License

Mixed Circuits

They are those that have both serial and parallel elements.

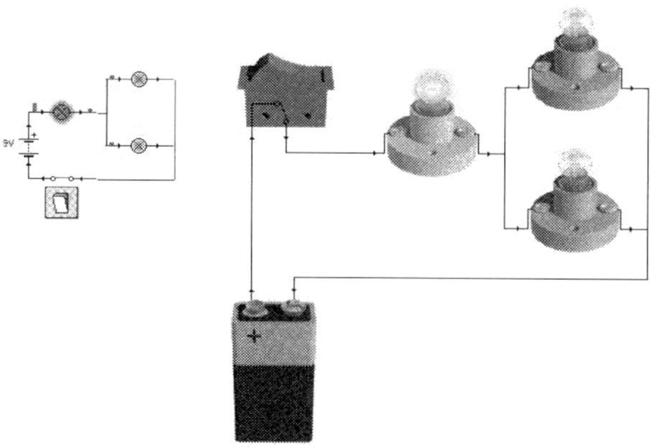

Figure 6: Mixed. Made by www.edu.xunta.es. CC-BY-SA License

Open circuit

For there to be current in a circuit, it must have a path through which it passes without interruption, the road must be closed. We say that a circuit is **open** when one of the parts opens, thus interrupting the flow of current. The flow of electrons stops this way. A circuit can be opened due to a fault, break or disconnection in one of its components: blown fuses, blown resistors, broken cables, loose connections... all this is often easy to detect with the naked eye.

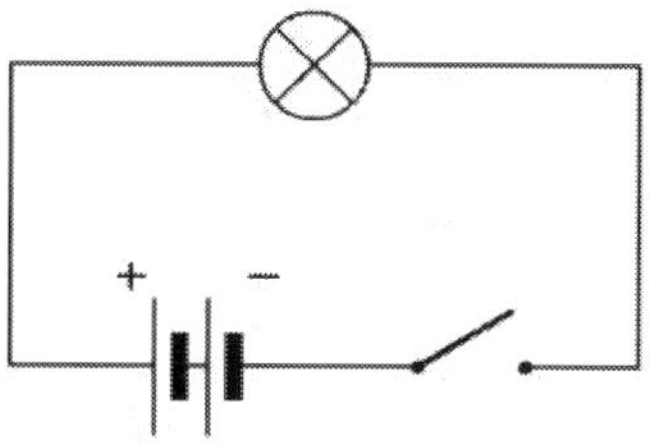

Short circuit

A short circuit occurs when the two current polarities are connected without any resistance, which causes the current to increase disproportionately. This occurs when the current has no resistance element to pass through and may cause damage or malfunction of the circuit.

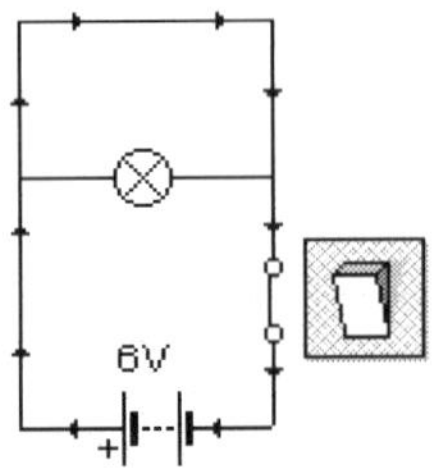

Figure 7: Made by www.edu.xunta.es. CC-BY-SA License

In the previous image we see that the current polarities meet in the top line, since there is nothing that offers resistance or reception.

Digital electronics and analog electronics

We can divide electronics into two main types: digital and analog.

Analog Electronics

It is the one that presents voltages and measurements that are continuous and not discrete, that is to say, that cover the whole scale of measurements in a progressive way. We are going to represent the voltage values in an analog signal:

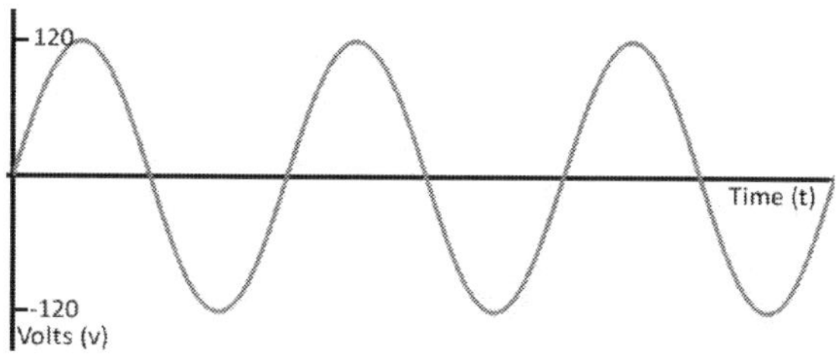

We see that the values do not go from 0 to 1 directly, but to reach 1, the signal goes through all the intermediate values, that is: 0.1, 0.2, 0.3... etcetera. But even to reach 0.1 you must first go through the values 0.01, 0.02, and so on. We will see practical examples of this, but first, by way of comparison, we must define digital electronics.

Digital Electronics

They are discrete voltages, not continuous, and have a defined value. The signal can go from 0 to 1 without intermediate values.

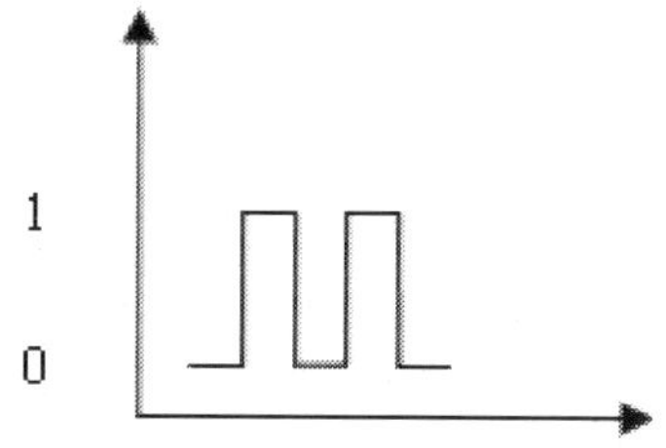

Let's look at an example of an old radio knob or dial. This is the use of analog electronics because the values are continuous and progressive. Now let's look at the example of the volume button on a mobile phone or cell phone. When we press the button, the values increase from 0 to 1, from 2 to 3, and so on. They do not increase progressively but the values are interpreted by the digital circuits as absolute and discrete.

We see that analogue electronics have values that are neither limited nor defined, whereas in digital electronics we find the opposite case. All signals are delimited to give certain values.

Digital electronics works with binary values: 1 and 0. This allows electrical signals to be transformed into data bits. The most commonly used components in digital electronics are diodes and transistors, which allow the creation of logical gates, thus creating a system capable of interpreting data and creating instructions, actions and processes from them.

Electronic Components

We will begin by identifying and defining the electronic components belonging to this branch. You can find them on most existing circuits and devices.

Resistance

It is a piece that opposes the flow of current. We use it to regulate and distribute the current flow according to the needs of each element of the circuit. Without a resistor, many components would be burned or damaged because the voltage or current is too high. That's why we use a resistor to send the right amount of voltage to the components.

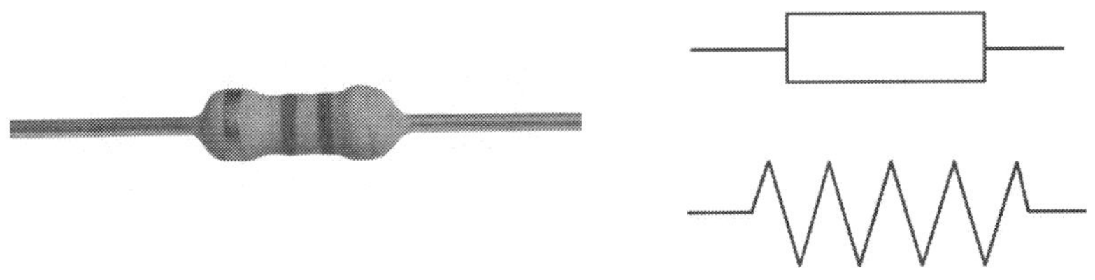

Figure 8: Resistance

Resistances have a certain value, and this is measured in Ohms. To know the value of each resistance without using mathematical formulas, we can observe the color code that these have. You can find color tables of resistances anywhere on the internet.

Capacitor or condenser

It is a piece that accumulates voltage, and releases it when fully charged. They can function as filters to reduce the noise of electrical signals. They can be electrolytic, which have defined negative and positive poles, and ceramic, which have no defined polarities. The unit of measurement they use is the Farads.

Figure 9: Electrolytic capacitor (right) and its symbol (left)

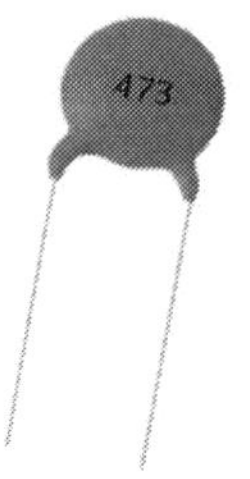

Figure 10: Ceramic capacitor

Coil or inductor

Coils are capable of generating a magnetic field when electricity is passed through them. They are made by a coil of enameled copper wire wound around a core, which may be air or some ferrous material. Its unit of measurement is the Henry (H). The coils act as opponents to sudden current changes, therefore they are used to give a more stable and controlled current luxury.

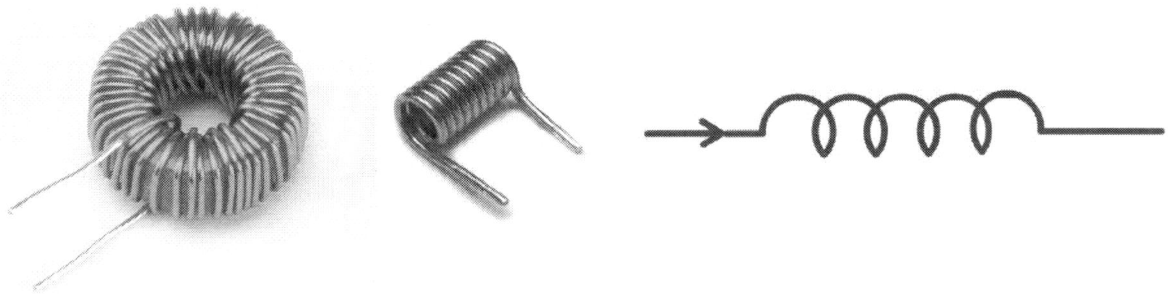

Figure 11: Coils (left) and their symbol (right)

Diode

The diode is a component that allows the current to pass only in one direction, and in the other direction blocks it. It consists of an anode (positive) and a cathode (negative). The diodes will always allow the flow of current from the anode to the cathode and not the other way around. One of the main functions that a diode can do is to protect the circuits from polarity reversal, which could cause damage to the components. There are several types of diodes: LED (light-emitting diode), rectifier diode, rectifier bridge and zener diode.

LED Diode

LEDs emit light when current passes through them.

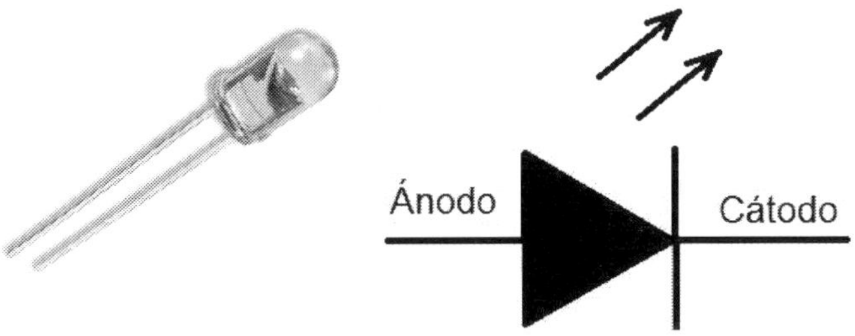

Rectifier diode

They are used in voltage sources to convert alternating current into direct current.

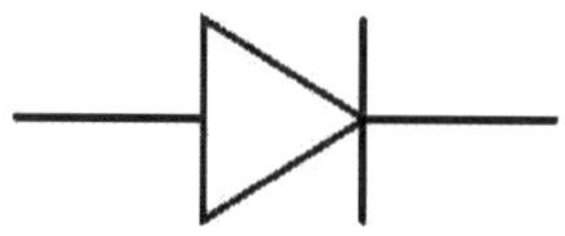

Bridge rectifier

They are used to stabilize an alternating current flow and convert it into a rectified wave, allowing this wave to be converted later into a direct current wave, using other elements.

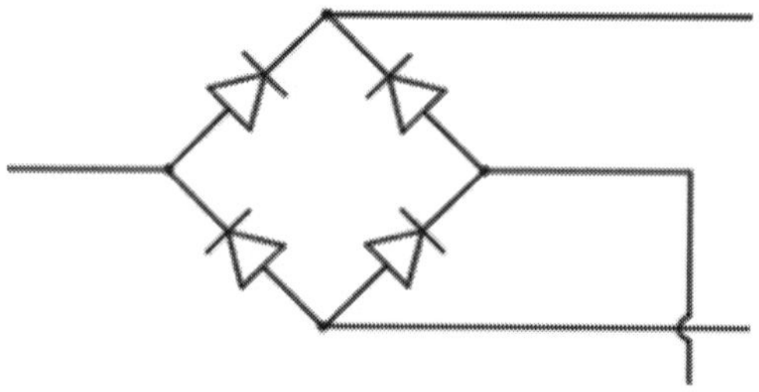

Zener diode

They are used to maintain a fixed and stable voltage, thus protecting the circuits from voltage spikes.

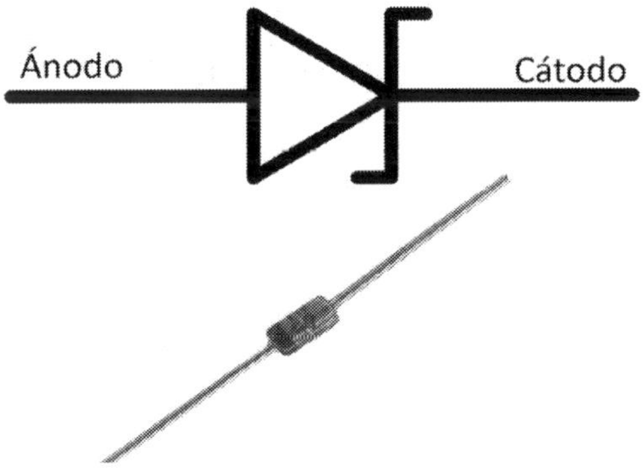

Transistor

It is a component of semiconductor materials that has revolutionized electronics and enabled enormous technological advances. It consists of three pins: emitter, collector and base. It has several functions. The first is to act as a switch, allowing or blocking electrical signals from a small control signal. It can be used as a logic gateway in digital electronics, since its components allow signals to be converted into logical digital language, which is why integrated circuits are made up of transistors. Analog electronics can act as a signal amplifier.

Tiristor

They are generally used as electronic switches. They are useful in power control and switching jobs. They have a very long wear life and do not wear easily.

Fuse

It is used to protect electronic and electrical components, allowing the passage of current while it is within acceptable limits. It consists of a metal filament designed to burn when current levels exceed the accepted limit. When burned, the filament no longer conducts current to the rest of the circuit, thus protecting it from high currents.

Switches

They are components that conduct, interrupt or allow the passage of electric current.

Printed Circuit Board

A plate or sheet of insulating material that has conductive lines on one or both sides. Along these lines the different electronic components of the circuit, usually soldered, adhere. Nowadays it is the most used method in the world for the connection and assembly of electronic circuits.

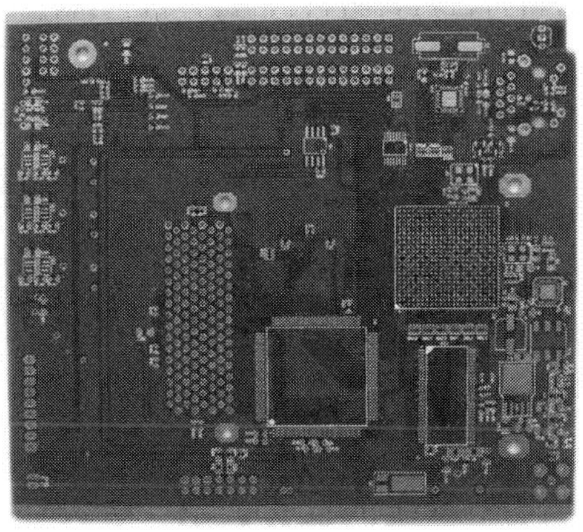

Batteries

They are an important source of direct current. They consist of chemical reactions inside them to accumulate electrical charges in their two terminals, which produces a stable voltage. The sides of the batteries are called anode (formed by a copper bar) and cathode (formed by a zinc bar), we also find the electrolyte, a liquid sulfuric acid diluted in water that tries to dilute both materials to produce a positive or negative charge.

Transformer

They consist of two coils wound around a core. It allows to increase or reduce the voltage of an alternating current flow, conserving more or less the same power, thanks to a process called electromagnetic induction.

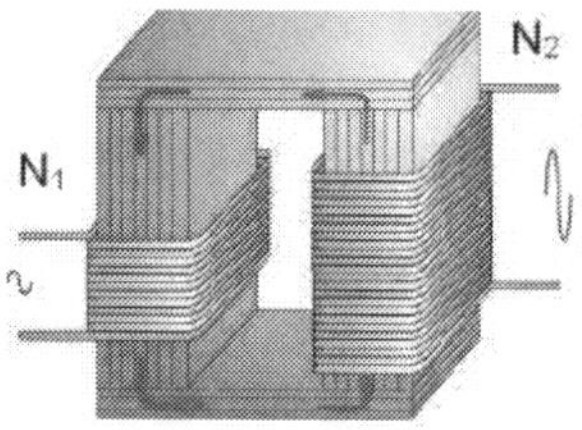

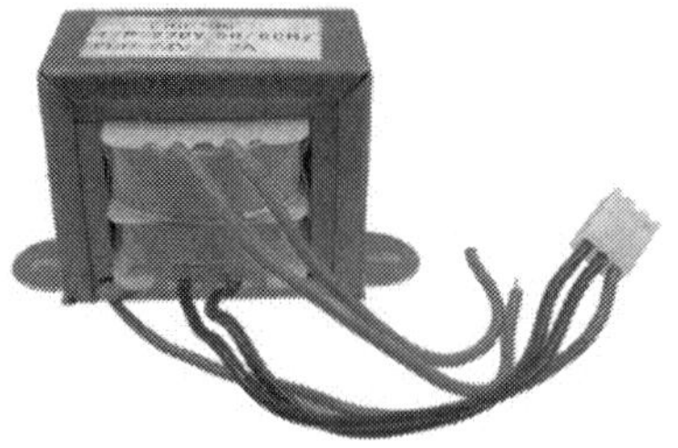

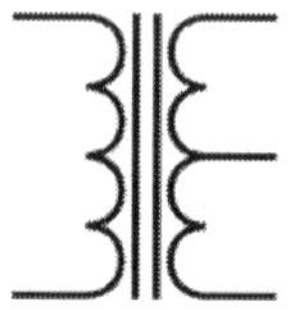

Integrated Circuits

They are widely used components in modern electronics, and have enabled huge advances in technology. They are circuits that can integrate hundreds to millions of microscopically sized transistors. They can perform numerous functions, perform complex calculations and process digital information. Thanks to integrated circuits there are computer processors, advanced scientific equipment, mobile phones, and almost every electronic device that exists today.

Tools to get you started

Multimeter

It is a measuring instrument widely used in electronics. It is capable of measuring practically all the parameters and frequencies existing in the circuits: voltage, resistance, continuity, current... among others. If you want to start practicing electronics, it is essential to have one of these. Each part of a digital multimeter is detailed below.

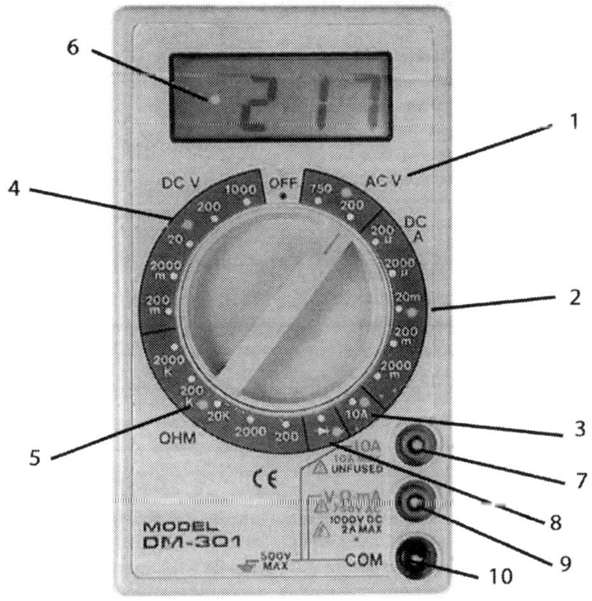

1: Voltage measurement in Alternating current (volts).
2: Measurement of current in direct current (amps)
3: High current in Direct Current (amps)
4: Voltage measurement in Direct current (volts)
5: Resistance measurement (ohms)
6: LCD screen
7: Terminal to measure up to 10 amps in direct current.

8: Diode measurement.
9: Red probe terminal (positive).
10: Black probe terminal (negative).

The multimeter will help you not only to measure but also to detect and repair possible faults and breakages in the circuits.

Oscilloscope

Measures electrical signals to observe the frequency, period, amplitude, noise, and behavior of a wave. All this can be displayed graphically in real time on the device screen.

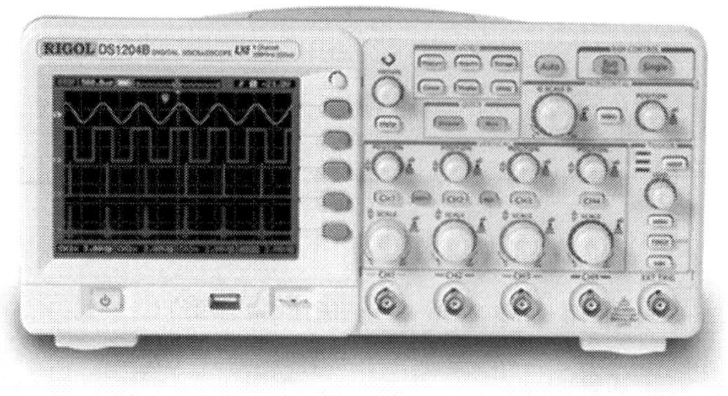

Welding station

Serves to join components by the application of heat and welding. Consists of a soldering or cautín pencil for soldering or desoldering elements using a precise tip. It also has a heat gun for welding or desoldering surface or heat shrinkable elements. The element used for solad is generally tin.

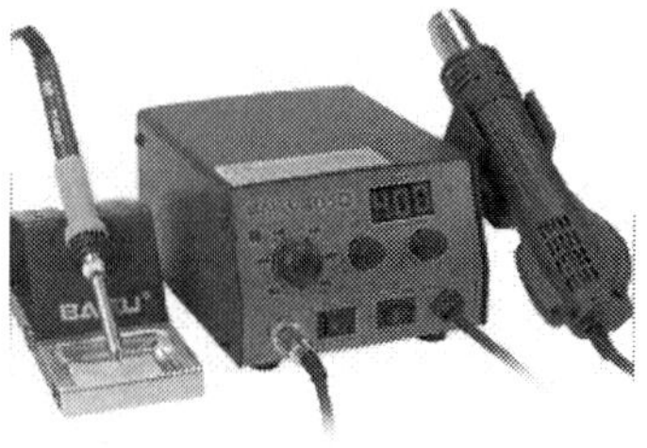

Protoboard Board

The connection board or protoboard is a board that allows us to assemble and disassemble electrical circuits with ease, in order to put into practice knowledge, experiment, or develop projects. All this can be done without the need for welding and with few or no tools. There are different sizes and shapes.

Screwdrivers

It is advisable to have a kit of screwdrivers with different shapes, can be cross, flat, with star tips, hexagonal, among others.

Tweezers

They can be flat-tipped, tapered-tipped, side-cut, or wire stripper.

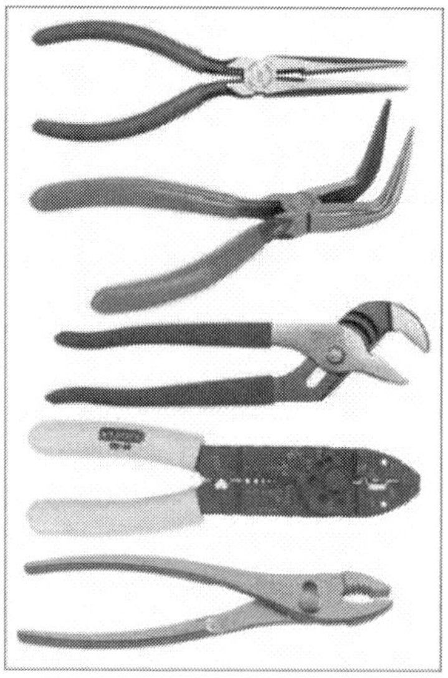

Electronic or electrical diagram

It's a graphical representation of a circuit. All components represented by signs are shown there, along with the lines representing the path of the conductors, the names and numbers of each component and the earthing connections. You will be able to identify the symbols of most of the components by reading, in previous lessons of this book, the illustrations of each component, which also include its symbol. We have included only the most basic ones, in internet you can find the complete lists of all the existing symbols in case you find one you don't know.

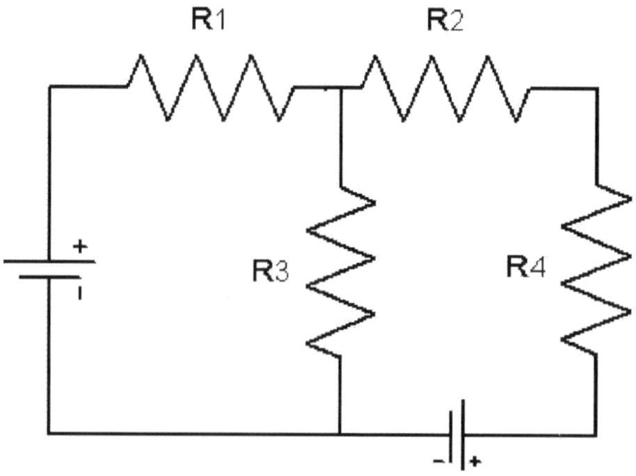

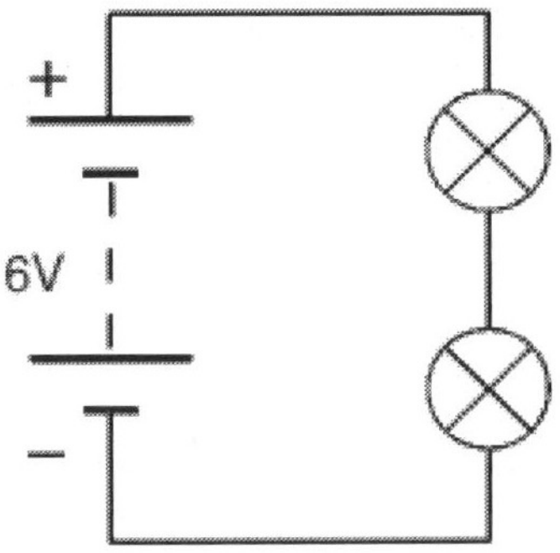

Figure 13: Circuit diagram with two light bulbs

Arduino

Arduino is an open-source electronic testing, creation and practice platform. Both hardware and software are free, and both are easy to use for those who want to get started in electronics. It allows you to create, modify small computers and circuits based on a motherboard provided by Arduino, to which you can add many electronic components that we have already seen in the chapter on components.

Figure 14: Arduino board with its inputs and components.

This board allows you not only to connect analog circuits, but also to program instructions to the integrated circuit using a programming language from a computer.

The different peripherals (cameras, sensors, motors, switches, leds...), are connected directly and send their connections to the main circuit, thus allowing you to manage them to your liking. The board also has an output interface where you can connect screens or speakers, also other boards or additional controllers.

The number of projects you can do with Arduino is very large. You can build everything from robots, voice-controlled devices, temperature and motion sensors, digital clocks, to anything your imagination can design. We recommend that you purchase an Arduino Kit that includes components, protoboard, board, and startup manual, so you'll have everything you need to get started in practice and it's very accessible. Once you've finished an internship, you'll be able to assemble another project and reuse it infinitely. In the Arduino starter kits you will find protoboards, resistors, capacitors, diodes, switches, LCD screens, LEDs, connecting cables, batteries and various components necessary for the system.

We hope that this little book has awakened in you the desire to know more about electronics. Many times, the first steps are the most important.

Printed in Great Britain
by Amazon

32945137R00019